AutoCAD
and Its Applications
BASICS

by

Terence M. Shumaker
Faculty Emeritus
Former Chairperson
Drafting Technology
Autodesk Premier Training Center
Clackamas Community College, Oregon City, Oregon

David A. Madsen
President, Madsen Designs Inc.
Faculty Emeritus, Former Department Chairperson Drafting Technology
Autodesk Premier Training Center
Clackamas Community College, Oregon City, Oregon
Director Emeritus, American Design Drafting Association

David P. Madsen
Vice President, Madsen Designs Inc.
Computer-Aided Design and Drafting Consultant and Educator
Autodesk Developer Network Member
American Design Drafting Association Member

2011

Publisher
The Goodheart-Willcox Company, Inc.
Tinley Park, Illinois
www.g-w.com

Library of Congress Catalog Card Number 2010009066

ISBN 978-1-60525-328-2

1 2 3 4 5 6 7 8 9 – 11 – 15 14 13 12 11 10

The Goodheart-Willcox Company, Inc. Brand Disclaimer: Brand names, company names, and illustrations for products and services included in this text are provided for educational purposes only and do not represent or imply endorsement or recommendation by the author or the publisher.

The Goodheart-Willcox Company, Inc. Safety Notice: The reader is expressly advised to carefully read, understand, and apply all safety precautions and warnings described in this book or that might also be indicated in undertaking the activities and exercises described herein to minimize risk of personal injury or injury to others. Common sense and good judgment should also be exercised and applied to help avoid all potential hazards. The reader should always refer to the appropriate manufacturer's technical information, directions, and recommendations; then proceed with care to follow specific equipment operating instructions. The reader should understand these notices and cautions are not exhaustive.

The publisher makes no warranty or representation whatsoever, either expressed or implied, including but not limited to equipment, procedures, and applications described or referred to herein, their quality, performance, merchantability, or fitness for a particular purpose. The publisher assumes no responsibility for any changes, errors, or omissions in this book. The publisher specifically disclaims any liability whatsoever, including any direct, indirect, incidental, consequential, special, or exemplary damages resulting, in whole or in part, from the reader's use or reliance upon the information, instructions, procedures, warnings, cautions, applications, or other matter contained in this book. The publisher assumes no responsibility for the activities of the reader.

Cover Source: Shutterstock (rf)

Library of Congress Cataloging-in-Publication Data

Shumaker, Terence M.
     AutoCAD and Its Applications. Basics 2011 / by Terence M.
Shumaker, David A. Madsen, David P. Madsen. – 18th ed.
     p. cm.

     Includes bibliographical references and index.
     ISBN 978-1-60525-328-2
     1. Computer graphics. 2. AutoCAD. I. Madsen, David A. II.
Madsen, David P. III. Title.

T385.S461466 2011
     620'.00420285536--dc22   2010009066

Introduction

AutoCAD and Its Applications—Basics is a textbook providing complete instruction in mastering fundamental AutoCAD® 2011 tools and drawing techniques. Typical applications of AutoCAD are presented with basic drafting and design concepts. The topics are covered in an easy-to-understand sequence and progress in a way that allows you to become comfortable with the tools as your knowledge builds from one chapter to the next. *AutoCAD and Its Applications—Basics* offers the following features:

- Step-by-step use of AutoCAD tools
- In-depth explanations of how and why tools function as they do
- Extensive use of font changes to specify certain meanings
- Examples and descriptions of industry practices and standards
- Screen captures of AutoCAD features and functions
- Professional tips explaining how to use AutoCAD effectively and efficiently
- More than 280 exercises to reinforce the chapter topics and build on previously learned material
- Chapter tests for review of tools and key AutoCAD concepts
- Practice questions and problems for Autodesk's AutoCAD Certified Associate and AutoCAD Certified Professional certification exams
- A large selection of drafting problems supplementing each chapter

With *AutoCAD and Its Applications—Basics*, you learn AutoCAD tools and become acquainted with information in other areas:

- Preliminary planning and sketches
- Drawing geometric shapes and constructions
- Parametric drawing techniques
- Special editing operations that increase productivity
- Placing text and tables according to accepted industry practices
- Making multiview drawings (orthographic projection)
- Dimensioning techniques and practices, based on accepted standards
- Drawing section views and designing graphic patterns
- Creating shapes and symbols
- Creating and managing symbol libraries
- Plotting and printing drawings

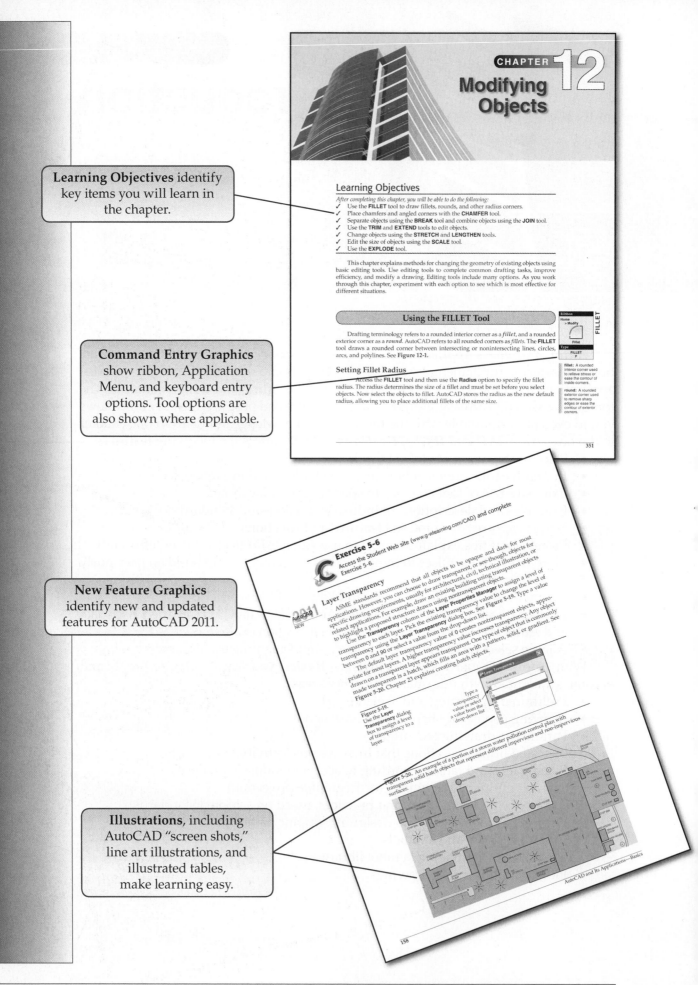

Learning Objectives identify key items you will learn in the chapter.

Command Entry Graphics show ribbon, Application Menu, and keyboard entry options. Tool options are also shown where applicable.

New Feature Graphics identify new and updated features for AutoCAD 2011.

Illustrations, including AutoCAD "screen shots," line art illustrations, and illustrated tables, make learning easy.